A 5G, ARTIFICIAL INTELLIGENCE

&

The Future of Business in Africa

Dan Maxwell, Jr

Why this, why now?

In 2015, I attended an educational technology conference about Africa in Addis Ababa, Ethiopia. There were at least 1,500 attendants at this conference.

It was a well organized conference. It provided lots fo value to participants. Everything went on without a hitch.

But all through the days at that concerence, I kept asking myself one question:

How could this event NOT have been organized by at African; dare I even say a black African – or person of African descent?

That's right, a conference about educational technology in Africa for Africans was organized by a White German lady.

This lady, whom I'll call maria, saw an opportunity and made the most of it.

Like Maria did with the hosting of an educational technology conference about Africa, many other Europeans and non Africans have done, and continue to do in other industries and sectors.

Here in Liberia where I'm currently based, propoablly the number one entertainment promoter in the country is of Syrian/Lebanese descent.

Now, let's be clear. In a globalized world such as ours, I have no problem with a citizen of one country starting a business and succeeding in another country. As a matter of fact, much my business is done with clients in countries outside of Africa.

What I will like to see however, is more black Africans getting involved in the continent's business landscape as the opportunities are enormous.

Whether you consider yourself an African, or of African descent living on the continent on in the Diaspora; or you are a non African with an interest in Africa's growth and development, the book is for you.

Artificial Intelligence (AI) and 5G technology (5G) sound like very complex and intimidating concepts, which is why when people even think of combining them, it feels like they're about to jump from the frying pan and into a fire.

You see, as complex as these subjects may seem, today, if you're going to thrive, or make a big impact in the world of business, you you will need need to have a basic understanding of them at the very least.

It is no longer a matter of choice. It is a matter of necessity.

Why is it now a matter of necessity?

Well, when you look around, you can quickly see that of all emerging technological fields, AI and 5G will have the greatest impact on the world in the coming decade.

Even more important is that these technologies affect the larger society.

With this book, I will help you cut through the noise so you can gain an understanding of how AI and 5G are combining to help build a brighter future for the globe, but more specifically, for Africa, a continent of amazing, extraordinary people with great potential.

From education to healthcare; transport and logistics to banking and finance; from agriculture to marketing, merging AI and 5G has the potential to revolutionize how the people of an entire continent live and work.

I wrote this book for the non techy person who has not had the time to really understand these two major technologies – AI & 5G are all about; and what they could mean for the ever expanding African business landscape.

Whether you currently own or run a business on the continent, or you're looking to start one, I believe you will find ideas, insights and tools that could prove useful for you.

There are some who would argue that both 5G and Artificial Intelligence are bad for humanity. If you're one of those, this book is not for you. Simple.

I do not seek to ponder on philosophical questions regarding whether machines can have a soul or not. Instead, it will help you understand how AI is going to increase efficiency and revolutionize how you live and work to create an all-round better world free of poverty and unnecessary suffering.

I know enough to know that one can find evidence to support whatever the viewpoint they already have; and when it comes to topics that involve humanity where everyone has their own invidual experience, you won't find me trying to argue in hope of changing ones mind.

If, however you're the person who really has not yet got this whole 5G and AI thing, and are genuinely interstly in learning more especially in connection to doing business in Africa, then read on. After that, I encounrag to find other material that presents a different view to mine, and then make up your own mind.

Raising the Global African Consciousness

The African Continent is the most minerally endowed continent. Yet, she receives the least benefit of her natural resources. African people spread across the world, also face a similar fate.

On the African Continent.

In the Americas.

In Europe.

In Asia.

People of African descent, most especially Black Africans have continued to be victims of inequality in Economics, Healthcare delivery, Education, and Infrastructural Development.

It is easy to see how the world seems to have ganged up against black people all over the world. I mean even on our continent, in our respective countries, somehow, our economies have come to be dominated by everybody but ourselves.

But while we march and protest about why and how our lives matter; and continue to advocate for fairer treatment, we must stop and ask ourselves some fundamental questions:

1. How did we create this?
2. How are we creating this?
3. How did we show up, consciously or otherwise, that led the rest of the world to look down on us?
4. How are were being today that we must continually be fighting for fairer treatment and our rightful place in the world?
5. How are we being that we don't get credit for our inventions and innovations?
6. How are we being that we are the least represented in international organizations?
7. How can we begin to see the world in order to change the current reality?
8. How do we want to see our continent and its people 50 years from?
9. How do we want to be 200 years from now?
10. What can we begin to do today to begin creating the future we want?

These are questions that no one person can have the answers to. And answers may not be what we need.

These are questions we must live from if we're going to change the global African narrative.

I believe the future of the Global African Community lies in its leaders playing a more pivotal and direct role in the continent's advancement. For me, that begins with self-leadership; taking personal responsibility for where we are as individual countries on the continent; concerning ourselves with the welfare of the African people spread across the world; and being mindful in how we relate to ourselves and others around us.

By providing high-level support to leaders who serve the continent and its people, and by fostering a meaningful partnership with like-minded individuals who are driven by the vision of seeing Africa Optimized, I envision an Africa that takes care of its own without the need for further exploitation masked as Foreign Aid.

I believe genuine social Impact Investments are the better way to go for Africa's development. Investments that are not geared toward projects that reward mental and physical laziness, but rather encourages exponential innovations in the private and public sectors.

Having been victimized by corruption at least once in the judicial system, and seen firsthand how the lack of accountability and transparency leads many to suffer, I personally endeavor to use my skills in helping Raise the Global African Consciousness, while working to shape a more passionate, mindful, tolerant and more inclusive world; where a country's people – an organization's clients, employees, leaders, and stakeholders all feel accepted, appreciated, engaged, and valued.

To this end, I'd like to introduce the social enterprise - **Africa Optimized** indicating our desire to make the best and most effective use of the naturally occurring and human resources of Africa for the good of Africans and their descendants all over the world.

The Vision: an Africa where every man, woman, and child lives from their highest potential, enjoying the benefits of good governance, and an extraordinary quality of life.

The Mission: Equip and Empower 10,000 African Consciousness Ambassadors who will bring a new awareness to our community.

While anyone can become an African Consciousness Ambassador, our preference will be teachers as they have the most influence on the next generations Africans.

Our initial activity will be the research and development of curriculum for teachers of elementary and middles schoolers. The African

Consciousness Curriculum will be used to teach principles of **Personal Responsibility, Self Leadership,** mindfulness, self-care, peace building, innovation, and financial literacy to the current and coming generation of Africa's descendants.

100% of Royalties Received from the sale of this book goes towards supporting our mission of Raising the Global African Consciousness through the Non-Profit Organization – **Africa Optimized.**

Raising the Global African Consciousness

To learn more, kindly visit
www.AfricaOptimized.com

Part 1

Understanding 5G & AI

Chapter 1: What is 5G?

If you look around you, you'll see that the modern life you live is courtesy of the existence of digital technologies such as tablets, phones, televisions, and smart appliances, just to name a few.

For these devices to function properly, they must be able to receive and transmit vital data and information. The process that these devices use to transfer this information is called *cellular networking*, which is a process that occurs in two main forms.

The first involves *voice transmissions*, which is how you can make phone calls. The second involves *data packet transmissions*, which is what makes it possible to connect to the internet.

The term *cellular* comes from the fact that one fixed-location transceiver placed by a cellular company can wirelessly service a large geographical area known as a *cell*.

Unfortunately, these transceivers have certain limitations involving speed, data capacity, and latency, which can make life difficult. Solving the limitations and inherent problems attached to these receivers is what has led to the invention of 5G technology.

Simplified to the simplest terms possible, 5G is the fifth generation of cellular networking; it allows technology-driven devices such as smart fridges, smartphones, tablets, and other devices to connect to the internet at a much faster speed.

Each of the four previous generations of cellular network technologies got their descriptions from their speeds. The 3rd generation cellular network (3G) had speeds of approximately 3.1 Megabits per second (Mbps); the 4th generation cellular network (4G) had speeds of approximately 14Mbps. And now, 5th generation cellular network (5G) offers between 50Mbps to 1Gbps.

Having a 5G connection means you get faster speeds for surfing social media, playing games online, streaming media, and downloading TV shows and movies. You get to spend less time waiting for downloads 5G, because it about 64 times faster than 4G.

To put that in further context, with 5G, you could:

- Download an HD movie in just 6 seconds instead of 7 minutes
- Save 2 minutes and 20 seconds a day on waiting for social media websites
- Save more than three hours when downloading a Spotify library of up to 10,000 songs

Each successive generation of cellular technology offered drastically faster speeds of data transmission, thus enabling devices to accomplish tasks that were unaccomplishable before. 5G technology provides three primary advantages over the previous generations of technology.

There are 3 Primary Advantages of 5G. The first is in an area called the ***Internet of Things (IoT)***. IoT is a new field in which thousands or even millions of seemingly unrelated devices can connect and share information autonomously without requiring a human being to tell the machines to do it.

The second is in **low latency information transfer**, which makes it possible to transmit large amounts of data over tremendously long distances at almost zero time.

The third is allowing devices that are on the move to maintain great data transfer speeds, thereby enabling you to have greater mobility without worrying about losing connectivity.

The impact of these three advantages and how they relate to artificial intelligence will become clearer as we delve into later chapters of the book.

But first...

Chapter 2: what is Artificial Intelligence

If you are like most people, then your first encounter with the concept of artificial intelligence is in dystopic science fiction movies from Hollywood, many of which involve robots, "AI," kill their human overlords, and take over the planet.

In reality, the truth about artificial intelligence is radically different from the imaginative power of Hollywood scriptwriters.

Artificial intelligence described the simulation of human thought processes by machines. For decades, computer programmers have been trying to mimic thought processes in machines in the belief that once computers can learn to think like people, they will be able to accomplish tasks that are impossible for human beings.

There are three main types of artificial intelligence programs.

The first is *Artificial Narrow Intelligence (ANI)*. This type of AI focuses on accomplishing a single task very well, but it cannot understand what it is doing or why. The industries expected to feel the revolutionary impacts brought on by ANI include health, transport, education, and agriculture. That's because these industries have many mundane and repetitive tasks that are easy to automate –and are better off automated anyway—which is where ANI functions at its best.

The second is *Artificial General Intelligence (AGI)*. This type of AI tries to mimic human consciousness so that a machine can experience what it means to be human. That includes having emotional experiences, forming bonds, and developing goals that are independent of computer programming.

Much research is still being done around the idea of AGI. When people stay Artificial Intelligence is not good for humans, they're usually referring to AGI which has the potential to create robotic human companions that can serve as caretakers for children and the elderly; but could also crash a stock market, highjack a plane, or any other number of things.

Creating an Artificial General Intelligence (AGI) is the ultimate endpoint for many AI specialists. An AGI agent could be leveraged to tackle a myriad of the world's problems. For instance, you could introduce a problem to an AGI agent and the AGI could use <u>deep reinforcement learning</u> combined with its newly introduced emergent consciousness to make real-life decisions.

The third is ***Artificial Super Intelligence (ASI)*** This is the AI technology that most fiction writers and movies focus on the most. In truth, this area is the least understood and most theoretical.

ASI endeavors to create a type of intelligence that far exceeds what the human mind can comprehend, which, in essence, would mean that humans would have finally succeeded at creating a new type of "being." The noticeable progress realized in this field of AI is tremendously impressive, with many computer scientists meeting important milestones every day.

Two decades ago, Deep Blue, the artificial intelligence built by IBM, beat Gary Kasparov, the best chess player in the world. Today, AI programs can replace drivers as companies like Tesla launch self-driving cars. The potential applications for artificial intelligence are virtually limitless.

Let's talk briefly about how the combined forces of 5G and AI will change Africa:

Chapter 3: How Combining 5G and AI is Going to Change Africa

What could we achieve if we were to become more productive, efficient, effective in Africa?

I believe if we're to take these technologies seriously as Africans, and be intentional about harnessing it's potentials, individual African nations can significantly improve standard of living for their people.

The relationship between Artificial intelligence and 5G technology is symbiotic. Improvements made in AI are helping create a faster, more efficient global internet system, which is, in turn, helping create better Artificial intelligence programs.

In the field of computing, latency is the time it takes for data to begin transferring after a command to do so. 4G technology has a latency period of 50 milliseconds, while 5G has a latency period of 1 millisecond.

Most artificial intelligence programs connect to online data services called *the cloud*. For AI programs to get efficient access to cloud services, the latency period between data requests and their receival must be as short as possible. That is why 5G technology is so critical to making artificial intelligence programs feasible and efficient.

To put this into a real-life context, in an interview with the BBC, Jane Rygaard, who serves as the Nokia head of marketing, stated that 4G technology could not offer online support to autonomous vehicles.

When it comes to preventing a road accident, every millisecond counts and could be the critical difference that saves a life —human beings can make vital decisions in less than two milliseconds. In comparison, the 4G alternative of 50 milliseconds is grossly ineffective.

Another way in which 5G will help make artificial intelligence programs viable is in the deployment of the Internet of Things (IoT). A real-life example of an IoT system would be the creation of smart cities where pollution and traffic congestion become better managed by feeding data from thousands of cars simultaneously to allow for easier navigation of urban areas.

Nairobi has 4 million people, while Johannesburg and Lagos have populations of 1 million and 17 million, respectively. That means traffic is often a problem in these cities.

5G technology can support more than 100 times more devices than 4G technology. All experts agree that 4G technologies will experience severe strains as millions of IoT devices come online and start making vast data transfer requests on an already overloaded system. That means only 5G technologies can make IoT viable in any kind of meaningful way.

Sometimes, Artificial intelligence programs require heavy computing power, leading to overheating devices and rapid battery drainage. This issue is already causing problems for mobile phones and wearable devices that utilize rudimentary AI programs.

5G technology uses 90% less energy per bit of data transferred compared to 4G technology. That means devices on 5G can conserve energy, thus preventing overheating and loss of battery life. Longer battery life will help the people of Africa save money —especially on electrical bills— and deploy AI programs while on the go.

As already stated, it's not just 5G helping deploy better AI programs; the inverse is also true. Research has shown that many network operators are already implementing artificial intelligence programs to manage their 5G networks better.

Predictive AI modeling is helping operators know in advance which areas will experience network usage spikes or declines. Operators are also better able to predict network failures and repair them faster, leading to better customer relations and cost savings. That happens through root-cause diagnostic tools that help operators identify the cause of a network failure that may take humans hours or days to find.

Artificial Intelligence for IT Operators, also called AIOps, is a newly emerging field that is using machine learning to help network operators gain insights into how data is moving across their infrastructure. With this, they can build autonomous 5G networks that can self-regulate without human intervention to reallocate network resources to where they need to be the most.

Part 2

How 5G & AI Will Change the African Business Landscape

Chapter 4: The Impact 5G & AI Shall Have on The Education Sector In Africa

"Education is the most powerful weapon which you can use to change the world."

Nelson Mandela

As a continent, Africa has vast natural resources and a young population that is eager and willing to work. For those young people to be able to reach their full potential, the first thing they need to do is learn skills that'll help them take advantage of the opportunities around them.

Education is the key to shaping minds and hearts that will be able to transform a continent that is still grappling with unemployment, poverty, and corruption into a shining beacon of hope and progress in the world. Artificial intelligence and 5G can play a significant role in helping educators reach their students in ways not thought possible before.

According to the Brookings Institution's Center for Universal Education, millions of African children do not have access to education. To make matters worse, more than 61 million children who do attend school end up not having learned how to read, write, or perform basic calculations. Researchers concluded that the main causes of these issues are teacher absenteeism, neglect, and teachers who don't understand what they're teaching.

Luckily, these problems are easy to solve with the adoption of intelligent tutoring systems. Such systems can gauge how far a student has already progressed by conducting simple tests. In many ways, these systems have an intellect that's superior to humans because they cannot be absent as a teacher would be, and they don't get bored of teaching the same thing every day.

A tutoring system will not advance to the next topic of a course just to get through the syllabus. Instead, it will conduct objective tests that ensure a student learns what is necessary before moving forward. This guarantees that every child learns at his or her pace. Equally impressive is the fact that the learning process becomes more personalized as each child can move at a comfortable pass instead of having a standardized approach to teaching children whose potentials differ so significantly.

AI-powered tutoring programs can grade student tests, a task many teachers consider quite tedious. The system can even identify areas that each student is displaying deficiencies in and assign courses and homework that caters to these particular areas.

Many schools in Africa, whether public or private, are offering their students access to low-cost tablets that are enabling students to have access to educational software. AI programs can be incorporated into the tablets to help improve the learning experience by providing personalized feedback, adaptive learning, and data analysis.

From pilot projects, researchers have observed that students are more willing to take risks and fail because machines are not as judgmental as humans, and this improves the learning experience because children only learn by making mistakes.

In classrooms powered by AI, the role of teachers would be radically different from that of ordinary schools; AI technologies would reduce the workload, which would make a teacher's job all the more exciting. Schools that adopt this new way of learning will have a major advantage in terms of service delivery compared to relying on previous models of doing things.

Many talented, older students who don't have the money to attend university can still have an opportunity to learn. Entrepreneurial Africans who wish to share their knowledge with students who are willing to learn are constantly setting up online schools and classes.

Online classes can have hundreds or even thousands of students who can engage with a single teacher. Machine learning algorithms can then detect which areas students have not understood properly and pass them on to the teacher who can then rehash them. This simple change improves overall efficiency as the teacher can direct effort and time into teaching areas that affect most students.

To deploy this kind of technology, we would require that the teacher have access to massive data transfer capabilities online so they can receive massive rapid input from students. The current 4G technology cannot handle this kind of data capacity because it operates at a much slower speed. Only 5G —and whatever comes after it— can support this kind of demand without latency problems to provide online interactive learning.

With only a mobile phone, a student in Africa will have access to invaluable information at almost zero costs. Once a student completes a course, that student can even receive an online certificate that's easy to authenticate and print. The students who graduate can then look for jobs using the credentials gained.

A real-life example of this is students who learn through the Andela training program in Kenya, Ghana, Nigeria, and Uganda. These students have developed a reputation for being so talented that tech companies in Europe and the US head-hunt them.

In 2019, the South African Department of Basic Education found that AI can help improve the quality of education in the country. In a dramatic display, an artificial intelligence empowered robot called *Zora*, accompanied the education minister Angie Motshekga.

Together they announced the deployment of AI programs as teaching assistants to tutors and personal teachers to students trying to learn coding skills. They selected 50 schools that are now receiving the Ms. Zora learning program. The program's goal is help SA become a global leader in robotics and AI technology.

Chapter 5: The Healthcare Delivery Benefits 5G & AI Shall Have On the African Continent

The African continent has a population of 1.225 billion, which makes up 17% of the world, and yet, it only has 1% of healthcare expenditure and 3% of healthcare workers. That means there's a gross deficiency in access to basic health services, a fact that shows itself in the data.

The WHO estimates that 60% of HIV-AIDS cases are in Africa; 90% of malaria cases in the world are in Africa. Luckily, the numbers have been going down thanks to conscious efforts by relevant stakeholders. Infant mortality and cholera still plague the continent, which begs the question, "how can we solve these problems —which shouldn't exist anywhere in our modern world.

Hospitals can improve service delivery to patients by sharpening day-to-day operational activities to save time and money. This chapter will show how disease diagnosis, patient monitoring, data analysis, and patient care are all improvable throughout Africa by adopting AI and 5G.

Disease Identification

Disease identification in Africa is improving a lot, especially in the field of cancer detection. Cancer is an illness that's often treatable if caught early. Unfortunately, identifying cancer can sometimes be difficult even for trained oncologists.

Machine learning algorithms funded by the Bill and Melinda Gates foundation and offered to medical physicians for free are helping doctors find cancer with far higher degrees of accuracy. These programs study thousands of mammograms and other types of x-rays. After a human instructed them where they were right or wrong, they could identify tumors with greater accuracy than humans.

Cancer is the second leading cause of death in the world, with most deaths occurring in low and medium-income countries and continents such as Africa. Such deaths cause massive losses of intellectual capital but also financial burdens on people and health insurance companies. With cancer detected earlier, it's possible to reduce these losses significantly.

Cancer isn't the only disease that AI programs are helping to detect. New medical diagnostic tools can conduct tests on blood, tissue, and urine, and offer diagnosis free of human intervention. AI has helped diagnose roseola, chickenpox, influenza, and glandular fever, just to name a few. Adopting this technology reduces the reliance on medical doctors and helps hospitals lower their costs of operations.

Wearable devices

Another area that is helping doctors and patients monitor health better is in wearable devices. These tools can monitor heart rate, blood pressure, blood sugar, and temperature. The devices then send the data to a doctor for analysis or perhaps even AI programs that internalize the information.

Companies like Fitbit use AI diagnostics tools to predict heart attacks and other health risks that could be building up. For this information to be meaningful to you, it needs to be relayed to your doctor or AI tool in real-time, with the information sent back to you as quickly as possible. That is one of the reasons why fast internet isn't just a matter of comfort; it could be a matter of life and death.

While it is true that 4G can relay data quickly, the devices of the future will be sending far greater loads of data than current devices because the more data your doctor has, the better the odds of coming to a correct diagnosis. For this process to be efficient, we require way faster internet than what is currently in use.

Early prediction

The global coronavirus outbreak that started in 2019 took the medical world by storm. Unfortunately, very few people know that an AI epidemiologist sent one of the first warnings regarding the threat it posed. Before the WHO and other health organizations had thought of taking the outbreak seriously, this software was already aware of the gravity of the situation.

The program worked by scraping news sources and blogs whispers on online forums. It analyzed the information, and despite government attempts at suppressing the information, it was able to detect that a storm was gathering.

By looking at which health websites people are visiting, what they are searching for on their computers, and what they are posting on their social media accounts, it is possible to detect flu outbreaks long before they become noticeable to human doctors.

Machine Learning and Big Data

Machine learning and big data analytics work to sort through millions of seemingly useless data points to create predictive models of what is likely to happen in the future.

That means a hospital with this kind of data could stock up on key medical supplies before other hospitals have a chance to take a run at them. Medical supply manufacturers can also be better able to foresee market demand spikes or slowdowns long before their competitors and adapt accordingly.

During the early phases of the coronavirus outbreak, manufacturers that had stockpiled ample respirators were able to sell them for high-profit margins. This shows that predictive modeling of disease outbreaks can help the healthcare industry improve, become more efficient, all the while saving money.

Better Hospital Management

Hospital management can also run better by ensuring that hospital command centers know which drugs are going to run out soon, which members of staff are performing poorly, and identifying potential areas of misdemeanors, etc.

Hospitals in Africa tend to have low cash-flow; they are therefore unable to stock up on sufficient drugs, leading to unexpected shortages. By predicting which drugs they're low on, these hospitals can always ensure that they never run out of stock.

Ambulance dispatching and patient prioritization is a task that can be difficult for a human to manage. Machines, on the other hand, can make these types of assessments at very rapid rates with far greater levels of accuracy as long as the information fed into the machine —by a human— is accurate.

Medical data on bills, scans, and prescriptions, all usually stored habitually, will now serve critical functions of helping hospitals deliver better care for patients. Patients will be able to make customer queries online and have their questions answered by qualified professionals. AI programs will also be able to schedule patient appointments better so that hospitals can reduce human clogging in their waiting areas. All this will serve to improve customer relations as people feel that their caregivers genuinely care about them.

Doctors in Africa today have the option of consulting their counterparts in countries like the US and Switzerland. They can do this through online medical forums or the Skype platform. Doctors today can also perform surgeries on their patients from very far away countries using real-time technology.

For these technologies to work, internet speeds need to be very high and capable of carrying high-quality images thousands of miles away at almost breathtaking speeds because one single mistake and a patient will lose their life.

In 2003, two South African doctors, Dr. John Sargent and Dr. Ernest Darkoh created a company called *Broadreach*. The company crunches data from 100 unique data points. It then aggregates it and determines the allocation of scarce medical resources for clinics in the country, which doctors are most effective, whether patients are taking their medicine, and which patients need extra attention.

The company then makes operational and staffing recommendations to hospitals to help them run better. Doctors have found the company's application to be so useful that of the doctors who have it on their phones, the number who used it went from 3% in April of 2019 to 65% within five months.

Chapter 6: How AI and 5G Will Make Transport and Logistics More Efficient In Africa

In recent years, there's been intense media interest in the field of autonomous vehicles powered by AI. The field has proven itself more than mere hype, especially since companies like GM, Mercedes, and Tesla have all launched their versions of self-driving cars. Estimates show that self-driving technology will lead to savings that amount to billions of dollars because they will reduce the number of accidents, eliminate human error, and reduce the costs of hiring drivers.

Unfortunately, the African continent will have to wait a while longer before this technology can become adaptable to its unique attributes. That is because Africa's roads do not have sufficient infrastructures like traffic lights, signposts, and well-marked roads. A lack of sufficient government funding in these areas has resulted in a less organized road transport network.

As things currently stand, self-driving cars can only work on the roads of 1st world countries like the US, Japan, and Europe. It will take more work before they become adaptable to the rest of the world.

These shortcomings do not mean that AI and 5G technologies cannot be beneficial to the African transport and logistics sector.

Hundreds of African companies focus on the transport of goods both within national borders and transnationally across the continent. These companies rely on large trucks that can haul as much as 25 tons per trip. That means each trucking haul could be worth millions of dollars. Therefore, any delay in delivery causes inefficiency and losses. This problem has already found a solution in the form of route optimization technology backed by AI.

New route planning systems can organize truck schedules to avoid peak rush hour times and traffic congestion caused by accidents. They can also coordinate schedules to avoid overburdened loading dock times at the main offices. The traffic data used to make these calculations can be sourced from Google maps as well as other private companies like Mapbox.com that collects and sells traffic data to over 600 million customers a month. In a country like Botswana, the government's Roads Ministry collects traffic data through physical car counters; the data then becomes available for use by the public.

Trucking companies are also able to reduce the number of accidents that their drivers become engaged in through machine learning algorithms that can sift through mountains of data to identify risk factors that are difficult to identify for a human being. These companies can then use this information to chart different routes and driving schedules. By assessing risk factors, they can also improve driver education.

When it comes to HR management, AI systems can help eliminate human bias in identifying which drivers are doing a good job and which ones aren't. An analysis of how long it takes drivers to deliver a haul as well as GPS data showing which drivers are exceeding speed limits can offer a glimpse into which drivers are performing well.

Truckers aren't the only people who can benefit from the AI revolution. Airline companies can use smart AI assistants to respond to customer inquiries on ticket availability, price, and flight schedules. Having such a system in place can help improve customer relations because they do not have to be put on hold before they can speak to an airline employee.

AI software can also help customers find their luggage at airports, which happens through tagging luggage with IoT devices and then using facial recognition software to match luggage with their owners. Internet of Things (IoT) requires very fast internet services. 4G technology may struggle to handle all the data transfers. This very fact makes 5G technology a necessity for all major airport facilities. Airports can also use facial recognition to identify potential security threats by studying facial features and behavior.

As is the case with truckers, AI can help airlines optimize their routes by sifting through millions of data points on weather patterns, customer inquiries, and booked flights to try and determine how best to avoid over or under-scheduled flights.

After fuel, the crew of an airline is the second-biggest cost. Organizing this crew efficiently can have a major impact on an airline's bottom line. A real-case example of AI scheduling crew schedules that might be helpful here is in the case of Kenya Airways, which was once known as the pride of Africa.

Crew failing to show up on time or outright mix-ups in scheduling led to major flight cancellations causing huge losses for the company. Perhaps the automation of such functions could have helped the company manage its employees more efficiently.

Airlines that want to keep their customers happy will also have to incorporate 5G technology into their planes. Flights often take hours; today, customers expect inflight access to the internet for entertainment as well as connectivity with their friends and families. In the future, airlines that will provide this consistently and at fast speeds will outpace their competitors.

The railway industry in Africa plays a critical role in the transport of goods and people across the continent. The coming revolution cannot leave it undisturbed.

AI systems using 5G technologies are helping in the real-time monitoring of queues at stations, efficient passenger and load distribution, and the monitoring of the performance of train technical performance.

South Africa's transport system has started to transition into a brighter future, especially because of a company called Intelligence Transport System that is helping public service vehicles plan their routes more efficiently. By connecting customers directly to PSVs, the buses can plan their trips from door to door. This, combined with integrated traffic data, means these buses are operating at greater levels of efficiency when compared to their chaotic competitors.

Kenya's ride-hailing company *Little* has also integrated AI into their operations to predict customer demand spikes as well as evaluate driver performance.

Chapter 7: How AI & 5G Will Help The Hospitality Industry Meet Customers Needs Better

Estimates show that the hospitality industry in Africa is worth an impressive $194 billion. Both local and international tourisms play a critical role in Africa's economy. The hospitality industry in Africa includes hotels, casinos, restaurants, and entertainment events.

Imagine if you were a guest in a foreign country having traveled for business and staying at a hotel. If the hotel had an overloaded Wi-Fi network, you could not be able to perform simple things like make video calls to your family back home, download work files or access online entertainment. That would make you say you shall never stay at that hotel ever again. That is why it is so important for hotels to incorporate 5G technology to keep up with consumer demands or else lose clients to competitors.

5G technology also allows hotels to incorporate new technologies that would not have been possible because of the operational limitations of 4G cellular networks. AI-backed chat-box applications with voice recognition software are already in deployment and allowing guests to make requests without having to speak to human employees at the hotel. Guests can request a taxi, change dinner reservations, adjust room temperature, place orders for special dishes, adjust room lighting, and manage itineraries.

Signature Lux Hotel in the Gauteng province, South Africa, offers intelligent voice-activated apps that can help guest check-in without any human involvement. Customers book their stay online and receive an automated email with their check-in details. The email also has a unique QR code that guests can use once they arrive at the hotel. They can then scan the QR code at a self-service key kiosk. From there, a guest has access to a room, with all payments made online, thereby effectively removing the human element from the whole process. Guests can also order meals and request other hotel services from the app.

In the past, regulatory burdens stifled casinos in Africa but governments have been loosening these restrictions of late, allowing the industry to blossom. Casinos can benefit from artificial intelligence by making it a part of their fraud detection apparatus. Casinos lose a lot of money to fraudulent clients who engage in card counting, slot machine manipulation, conspiracy, and chip switching.

AI-backed facial recognition software can identify players who have been previously flagged by other casinos for cheating. Many casinos share information on criminal elements because it is in all their interest to lower instances of fraud across the entire industry. Gambling addicts who have registered themselves onto self-exclusion lists are also identifiable through facial recognition software so that management can warn them not to engage the casino floor.

A lot of the fraud in casinos is by guests working in tandem with a casino employee. A single employee working for months or years can cause sustained losses over time that can bring down the entire business.

AI-backed software can identify instances of customers who are winning at above normal rates and then link this with employees who are present during these instances.

Sifting through this kind of data would require dozens of people, which would be an additional cost for the casino. Additionally, there's no guarantee that these extra employees won't be in on the fraud. It's impossible to bribe or deceive intelligent AI, which is why incorporating it would be tremendously beneficial for this industry.

AI can also help casinos create marketing campaigns targeted at specific customers. Facial recognition software can identify players who engage in high-value gaming and offer such customers loyalty reward points as a way of making them feel valued as customers. AI makes it possible for customers to interact with marketing ads. An analysis of customer engagement can help casinos figure out which ads are working on particular demographic points. All this information would require hours of human resources to compile. For an AI program, it could take less than a few seconds.

The AI and 5G revolution is not leaving behind the restaurants. Guests at restaurants sometimes feel pressured by waiters who have to go around dozens of different tables to collect orders. An artificial intelligence backed chat-box can help overburdened staff by providing customers enough time to make orders while the waiters focus on service delivery.

Customers can also take time to ask specialized questions on particular meals like the origin of the ingredients as well as the nutritional value and calorie amount in each meal. It might be difficult for a waiter to have all this information, but an AI can have all the information at the touch of a button.

AI-backed robots also have great potential as possible replacements for waiters who would transition to managerial roles. That would help hotels cut down on the costs of labor and remove the human element of making mistakes.

Nowadays, it is all too common to see a customer at a hotel complaining of a mix-up of his or her order with that of someone else's or getting the wrong order altogether.

AI eliminates all these kinds of mistakes. Telecommunications giant Vodacom has launched one of the world's first commercially available 5G networks in Lesotho. The company's Chief Technology Officer has stated that one of the primary transformations that will take place due to 5G technology is in the robotics industry of Lesotho. Perhaps restaurants may play a role in this transformation.

AI systems integrated into a restaurant's Stock-and-Point-Of-Sale software can help hotels understand their customer and stock-flow patterns better. Predictive modeling can help forecast when certain items might run out of stock, thus providing a chance to order them in advance.

Chapter 8: How AI and 5G Will Optimize the Manufacturing Process in Africa

Many experts widely believe that the manufacturing revolution that helped Asian tigers transform their economies to rival that of traditional 'first-world' countries is moving to Africa next.

It is no longer a secret that China's economy is slowing down. If Africa is going to take advantage of the tremendous opportunity that exists then, it must be willing to embrace AI and integrate 5G into manufacturing operations.

In 2015, Africa had an estimated manufacturing output of $500 billion; by 2025, analysts and experts predict that it'll well be over $1 trillion. Morocco manufactures plastic materials; Nigeria has a very robust cement industry, Madagascar makes biological fertilizer, and South Africa has the most industrialized economy on the continent, exporting everything from computer electronics to chemicals.

Manufacturing companies can benefit from the reliability and lack of latency that 5G offers. A lack of latency means an operator can control systems from anywhere in the factory, and the system will respond in real-time.

The manufacturing sector is one of the first industries that is enacting the Internet of Things (IoT) technology. As of right now, the main use is in monitoring and controlling the hundreds or thousands of machines that are operating simultaneously on a factory floor. For this to be viable, we need to replace the current 4G system with 5G, which has far higher data and unit connection capacity.

Factories use manufacturing lines where they divide work into specialized functions. A single failure in one area creates a backlog that stops all other areas. The failure could be due to technical problems or insufficient human resources allocated to a task.

An AI system that is monitoring movement on the manufacturing line can help create a more efficient system by identifying early signs of a build-up before it becomes a major problem. AI can then help deploy additional staff if the problem is an insufficient workforce or, if the issue is mechanical, it can help deploy technical experts.

More humanoid robots can also take on functions that human operators have on factory floors, something that would undoubtedly help reduce the number of employees and save businesses on payrolls. Robots don't get tired or sick; they just perform the tasks given to them; they follow their programming.

Some AI critics have pointed out that when robots replace humans in factories, it will create mass unemployment. Nothing could be further from the truth.

AI will eliminate certain low-paying jobs, that's for sure. But, it'll replace them with better higher-paying jobs, all while allowing businesses to make greater profits all round.

The idea of massive unemployment caused by AI is a myth propagated by those who have not studied the issue in greater detail. AI will create a world where everyone wins regardless of whether they are manufacturers, customers, or employees.

5G technology means that separate manufacturing units connected to one intelligent entity can have real-time access to data stored on a cloud service. A cloud service means separate parts do not have to store data individually, which can be expensive, but can instead have access to an online data storage hub. Only 5G can provide access to this information at speeds that make this approach viable.

An interesting use for AI in manufacturing is in quality control checking. Some areas of production are very sensitive and require 100% accuracy, failure to which could lead to the loss of lives. An example of this is in the manufacture of vehicle or airplane parts. Accuracy is why human quality inspectors have to check the various components for any potential flaws that could compromise the end product. Luckily, this process is now automatable, with machines (AI) being able to provide greater accuracy at lower costs.

It is now possible to use AI-backed machines with HD cameras that can see hundreds of times better than the human eye to find flaws in manufactured parts.

Finally, AI can help factories in the area of predictive maintenance. Businesses across the world lose billions of dollars to inefficiencies and delays caused by faulty machinery. Current maintenance operations rely on Planned Preventive Maintenance (PPM), which relies on factors like expected lifespan and scheduled checks. This approach is inefficient and runs the risk of over or under maintenance of an item.

AI-backed Preventive and Predictive Maintenance relies on real-time analytics derived from sensor data checking on asset performance, temperature, and weather exposure. The system then compares that information to historical data and tries to estimate when an item may need repair.

The ability to predict flaws and quickly identify where they are can help factories drastically lower costs. AI systems integrated into every part of the manufacturing process can identify declines in output and warn human operators of potential flaws in the process.

Despite these massive benefits, it is necessary to point out a major risk for manufacturers worried about protecting trade secrets and preventing cyber-attacks from rivals because 5G could create a security risk that is unique to the network.

5G technology can transfer massive amounts of data very fast. That means cyber-attacks can happen almost instantaneously. Manufacturing floors where all systems connect to a single hub are particularly vulnerable. That means better cybersecurity is a necessity to protect manufacturing facilities. Luckily AI-backed cybersecurity solutions are under development, and they will be able to react to intrusions at faster rates than a human could.

Chapter 9: How 5G and AI are Revolutionizing IT Services In Africa

IT services in Africa are an area that has seen tremendous growth that's almost comparable to that of 1st world countries. From website design and mobile development to eCommerce systems and game development, Africa has made enormous strides in being a global leader in the IT industry.

African IT businesses are also incorporating AI services into their platforms. An example of this would be a software called FinChatBot in South Africa, which was launched by Far Ventures LLC. The company builds chatbots for websites, which helps increase sales conversion rates, predict customer needs, and answer questions that potential clients may have.

Another impressive IT company incorporating AI is an Egyptian firm called Affectiva, which specializes in using facial expressions and human behavior to detect emotions. The company can then sell this data to—or share it with—the health and gaming companies. Affectiva has raised over $50 million for its innovative work.

Facebook is also working with Kenyan firm Cellulant to create an augmented reality feature for eCommerce platforms. This augmented reality feature will allow users to test out products like clothes before choosing to buy them, which will improve customer experiences and reduce the number of returns from unhappy customers who ended up disappointed by an order.

Artificial intelligence can help African website designers create platforms that can sift through data to determine how customers interact with websites. From this, they can figure out how to increase sales and change parts of their websites that are difficult for users to navigate.

Today, high-end virtual assistants have become almost indistinguishable from the real deal. This workforce is helping IT companies address customer needs without having to hire extra staff to talk to customers. Only when an AI fails does a real person step in to help out a customer.

Another problem currently affecting African mobile and web developers is the creation of interactive platforms that can sync up users with cloud services or servers. If a website can link up with online services without latency and downtime problems, then a developer can be comfortable with creating light versions of applications that can access necessary information online in real-time. 5G is fast enough to allow for this kind of service to be available.

IT companies launching games will also benefit from 5G technology because today, players are looking to be part of a global community. Gamers nowadays play with people from different nationalities online. For this technology to be viable in Africa, then both speed and data capacity must be up to the task, which is especially true today as games develop more sophisticated graphics requirements.

Chapter 10: 5G & AI and the Future of Telecommunications in Africa

Telecommunication giants like MTN in Nigeria and Vodacom in South Africa have revenues in the billions of dollars and are a source of pride for the people of Africa. Currently, there are only two African countries that have launched 5G: Lesotho and South Africa, both courtesy of Vodacom.

One way in which telecommunication firms in Africa can benefit from using AI is in calculating the Customer Lifetime Value (CLV) of each of their clients.

We calculate CLV by assessing a customer's spending habits and service utilizations to determine whether the customer will provide a loss or a profit over a long period. This calculation helps telecommunication firms segment their customers into one of three categories: profitable, almost profitable, or unprofitable. This can helps firms predict future cash flow patterns and determine how to create effective marketing strategies.

Calculating CLV is difficult because market conditions are always in a state of flux, which means it can be difficult for human beings to do this manually. From new taxes to competitor price changes, many variables impact the CLV; it is better to have an AI calculate this number than a human.

Another interesting area is AI-backed predictive analytics. This area relies on machine learning to help telecommunication firms determine usage spikes before they happen. Once a firm knows that a spike is coming, it can allocate network resources to particular regions at specific times.

Network failures due to overloading cost telecom firms a lot of money and lead to unhappy customers. Predictive analytics works by analyzing past historical data of network usage to build forecasts based on probability.

Predictive analytics also helps telecom firms manage their employees better because management has an insight into how many employees they will need at particular times so they can create work schedules based on this data.

It's an unfortunate reality that the telecommunications industry in Africa is rife with fraudulent activities by scammers trying to get at people's money. These scammers rely on fake profiles, SMS phishing, and illegal network access to cheat people. Telecoms are unable to root out these bad actors because sifting through millions of accounts to find a few hundred bad ones is next to impossible for a human.

An AI can identify these bad actors by finding patterns of behavior. An example is identifying certain keywords used by criminals in text messaging scams and then cross-checking this information with the sender's physical location and past behavior. That information can then become a red flag used to identify that something might be amiss, and a human operator should look into it.

Once telecoms in Africa roll out 5G fully, there will be minimal downtime, greater speeds, and better capacity. This enhanced capacity will allow these companies to lower their costs for their clients and reduce labor spending of facing network problems. That will lead to better quality for customers at lower fees and higher profits for the network providers.

Chapter 11: How 5G & AI Can Increase Food Production and Processing In Africa

Agriculture plays a huge role in Africa's economy. The World Economic Forum estimates that 24% of the continent's GDP is in agriculture, and 60% of Sub-Sahara's population engages in one or the other form of farming, even if it's on a small scale.

Africa also engages in a lot of food processing activities —this is the process of turning agricultural products into familiar forms of food. Examples of food processing operations include canning, pasteurization, freezing, fermentation, grinding, and smoking.

Despite Africa's tremendous strides in creating food security, estimates indicate that 20% of the continent's population suffers from chronic undernourishment. The only way to resolve this problem is to increase food production by utilizing technology that raises crop production for every unit of land. This chapter will show you how

Artificial intelligence and 5G can play a huge role in this endeavor.

Deep Learning

We have a subset of artificial intelligence called deep learning. This field focuses on trying to imitate the neural network of the human brain, the human organ responsible for the storage of memories and processing them.

The use of Deep learning has been beneficial use in the field of agriculture where it is helping create better seeds for farmers. A crop's ability to fight off infections, produce nutritional outputs, and generate high yields depends entirely on its genetic makeup.

Scientists are already working on increasing the output of sorghum, which is a staple food in many African countries that have arid conditions. Unfortunately, sorghum has over 33,000 genes, and identifying the gene responsible for individual functions is a time-consuming nightmare. That is where AI can come in and help sort through the information far quicker than humans. The AI does this by studying mountains of previously collected data on different types of sorghum and their genetic makeup. It then tries to identify patterns that then give scientists reasonable starting points as opposed to randomly guessing for 33,000 genes.

Better crop management for large scale farmers

AI can also help in crop management for large scale farmers. Corporation growing crops on thousands of acres often find it difficult to keep an eye on large areas of farmlands. Failure to check on crops can lead to massive losses when diseases start on one part of the farm then spread to the rest of the crops.

Crops are also prone to attacks by large swarms of indigenous, dangerous insects like the locust plague that moved from the Middle East and into East African countries like Kenya and Somalia, causing massive devastation in its wake.

To handle these risks effectively, farmers often hire manual laborers to check on the crops, a strategy that is time-consuming, inefficient, and expensive, which is why farmers are now turning to AI-powered robotics to solve the problem.

AI-powered drones can identify sick crops by reflecting near-infrared light on them. The multispectral images collected can then undergo analysis because healthy green plants reflect light differently than sickly plants that have started turning yellow. These machines can even detect changes in plant color before they are visible to the human eye. The AI can also distinguish weeds from genuine good crops to inform a farmer of the potential risk they pose.

The ability to identify individual weeds or crops that are sick has led to the emergence of a new way of farming called *precision agriculture. Precision agriculture* allows farmers to use drones to spray specific areas with agricultural treatments and determine which areas need additional water and fertilizer inputs.

A South African start-up called Aerobotics is using machine learning to process drone and satellite data to provide farmers with insights on crop health. The company has now raised more than $2 million from venture capitalist impressed with their progress.

Autonomous robots are also taking on roles that humans used to do. AI-backed robots can now check on crop output and collect the produce when it is harvest time. An initial investment in a robot can eliminate costs on labor for years to come.

Artificial intelligence is also helping with the study of localized weather patterns to predict crop output and therefore estimate price volatility. This information is proving very useful to traders in the agricultural futures market who have to buy crop produce weeks or even months in advance.

IoT devices can also be effective when integrated into greenhouses to boost crop yield. By monitoring conditions in the greenhouses —like temperature, light, and humidity— these devices then regulate conditions increasing UV light or opening air intake valves for greater circulation. IoT devices working in tandem with each other require 5G technology to guarantee zero downtime and latency.

Companies in food processing are already using AI to help sort inputs. Whether it's in the processing of fruits, coffee, wheat, cocoa, or tea into final goods, the crops received from a farm always come in different sizes and quality. It is often necessary to sort out the different qualities of crops because they produce different quality processed goods. Sorting through them by hand is a tedious process, but AI using HD cameras and Near Infra-Red (NIR) spectroscopy can do this work very easily.

Chapter 11: The Effect 5G and AI Shall Have on Banking and Financial Services In Africa

A feature that applies to most banks of the world is that everyone hates going to them. From long queues to tired attendants, the entire physical banking experience leaves a lot to be desired. Luckily AI and 5G could help improve the situation for an all-round more enjoyable experience.

The role of the banking industry in Africa is to act as a conduit of financial payments, to provide loans to businesses and individuals, to store peoples' savings, and to offer interest payments on those savings.

Thanks to technology, many of these activities no longer require customers to travel to banks for services. New emerging banks are now allowing people to open accounts from the comfort of their homes using mobile apps. People can save money on their mobile phones, and, based on their credit history, borrow personal or business loans.

Higher Security

One of the biggest impediments to online banking has been security concerns. A fraudster who gets access to your password could siphon your bank account of all its money. AI-powered biometrics are helping solve this problem by creating stronger security systems.

A combination of fingerprint scanning and facial recognition done from a conventional smartphone is enough to authenticate identity before the completion of a transaction. New artificial intelligence programs can now detect stress in voice patterns and facial expressions. That means even if someone holds you at gunpoint perform a transaction, there is a good chance the AI could detect that something is wrong and reject the transaction.

Efficiency (and investment advice)

Chatbots are also helping banks address customer needs without having to hire extra staff to monitor customer queries. Chatbots can handle basic operations like opening an account, transferring funds, or queries on loans. Chatbots can also serve as personalized financial advisors to help you determine which investments you should make based on your risk tolerances.

One of the biggest problems facing the banking industry is money laundering. Terrorists and drug cartels try to escape tracking by security organizations by merging the finances of legitimate businesses with their illegal activities as a way of 'cleaning' the money. Today, AI systems can analyze a potential client's online and offline data from emails, social media, IP addresses, and phone numbers to identify any kind of potential red flags. Banks linked to illegal activities suffer from public relations disasters, which is why it's often in their best interest to prevent these situations from emerging in the first place.

Tyme bank is a South African digital-only bank that is trying to eliminate any need for physical interactions between the bank and its customers. According to their Chief Data Scientist, the company is disrupting traditional models of banking by using AI and cloud computing to provide all ordinary banking services at nearly zero costs. This technology has allowed them to grow exponentially and gain over 600,000 customers within a matter of months.

Banks aren't the only firms in the financial sector that are profiting from the AI and 5G revolution. Financial trading firms have invested heavily in the coming technology revolution.

Traders in financial markets like stocks, indices, currencies, and commodities all rely heavily on very fast internet. These companies don't just invest in the best network providers; they also invest in having the best internet cables and physical locations. In the world of high-frequency trading, success or failure depends on being ahead of your competition by milliseconds. That means 5G isn't a matter of comfort for these firms; they need it to remain competitive.

The last decade of finance has seen strong performance from investment funds known as quants. These firms don't make investment decisions based on human actions; they use software that utilizes sophisticated AI system.

Financial firms build customized software that tracks technical indicators like trading volumes and moving averages to decide which investments to make. The speculative nature of these investments means that they often last for a few minutes or even seconds. The success of this approach means that Africa's financial firms must embrace this new way of doing business or risk becoming uncompetitive.

Africa has seen an increase in mobile loan companies in the last few years. These companies receive hundreds or thousands of requests for loans per day. As you can imagine, they can't go through each request individually. The ability to determine which clients are creditworthy is a matter of utmost importance for these companies.

Machine learning algorithms can sort through financial records to help identify patterns of which type of customers pay back loans and which ones don't, which can help them sort through the loan requests in a matter of seconds instead of hours or days.

Chapter 12: Effects of 5G and AI on the African Fashion and Design Industry

There is a high global demand for high-quality apparel, but the clothes business is a labor-intensive industry. That is why many multinational corporations have set up their manufacturing operations in Asia and Africa. Unfortunately, the conditions in these factories are often not ideal workspaces, leading to the use of the term 'sweatshops.'

The use of AI and 5G can help manufacturers alleviate the suffering of their employees by increasing output per person and eliminating overly tedious tasks.

Autonomous AI-backed robots can take over mundane tasks like sewing, printing, braiding, weaving, and dyeing. Human operators would then focus on managerial tasks leading to better working conditions and better pay. The businesses would save labor costs leading to high profits.

A critical part of apparel manufacture is quality control. After the manufacture of a shirt or pair of jeans, there has to be a human inspector who assesses the cloth to make sure it has no tears or misalignments. Integrating robotics, HD cameras, and machine learning algorithms can help automate these processes.

If an AI-backed machine sorts through thousands of clothes with a human operator informing it when it sorts correctly and when it makes a mistake, the AI can learn which mistakes to look for; eventually, its accuracy rate would surpass that of humans.

Many thought the idea of an artificial intelligence designing clothes was impossible because fashion feels like a very subjective, intuitive, and uniquely human experience that should be impossible for a machine to do. Once again, AI is proving skeptics wrong with the consistent launch of new AI fashion programs that are proving themselves up to the challenge of competing with human designers.

The AI works by scouring social media platforms to find fashion trends that are just emerging in popularity. The AI then distills the unique characteristics of these fashion trends and tries to emulate them to create something unique and beautiful. Nuanced details like the shape of a collar, the design of a button, and placement of pockets are all data points that AI can factor in to determine emerging fashion trends.

AI is also helping fashion retailers sell their merchandise at a faster rate. AI does this by studying customer's previous buying patterns to identify their tastes. With this information, it is possible to create marketing ads personalized and tailor-made to individual people or demographics of people.

Facebook ads allow clothes sellers to target their potential customers by very detailed categories ranging from location and financial capabilities to age and sex. That means if a marketer already knows what certain customers want based on past buying practices on their platforms, then their marketing campaigns can have a high turnover and save them money.

A Nigerian firm named Touchabl allows users to find out more about an image by touching parts of it. The app's main userbase is people hoping to find out where they can get clothes that they liked on friends or celebrities. The firm has already raised $21,000 and expects to invest more into AI so that their software can deliver better search results for their clients.

Chapter 13: How AI and 5G Will Improve Real Estate Development In Africa

Africa has a real estate industry that's booming at an exceptional rate. The main reason behind this boom is the fact that the population of Africa is young and growing at a rapid pace. Many of these people are educated and joining the middle-class, which means they can afford to live in nice modern homes.

One of the biggest problems impacting the real estate sector is project cost overruns. Project cost overruns occur due to design errors, changes in materials, poor management, and design flaws. A lot of these mistakes are a result of human bias as contractors try to make overly optimistic projections.

AI programs can use data on project materials, the project design and site location, and based on past projects, create a better estimate of how much the project will cost. The AI is unbiased and not influenced by human emotions or poor judgment, which means its estimates are more likely to be correct.

AI can also help real estate companies estimate demand in particular localities. As the old real estate saying goes, "location is everything"' choosing where to establish a building is the single most important decision that a real estate company can make. An area with insufficient demand can lead to substantial losses for a real estate developer.

AI can help determine the demand in an area by using big data analytics to assess migration patterns and determine whether a neighborhood is rising or declining based on keyword searches.

The collapse of the real estate bubble in 2007 led to the greatest financial crisis since the great depression. This bubble was a result of mortgage lending institutions giving money to people who were not creditworthy. AI can help sort bad credit risks from good ones by utilizing a data-driven approach that analyzes an individual's past financial patterns to determine their risk levels.

AI can also introduce autonomous machines that handle mundane, repetitive tasks on construction sites such as pouring concrete, welding, and wall demolition. That can free up human workers to focus more on nuanced and difficult tasks that machines cannot handle.

AI can also help reduce costs emerging from worker injuries on work sites. Cameras on construction sites can identify workers who are not wearing proper safety gear as well as areas on construction sites that are safety hazards for people.

Real estate is an area that most people associate with a lack of progress because how we build houses hasn't changed much in over 100 years. All this is changing with the advent of smart homes. Appliances in Smart homes connect to a network that's accessible remotely.

A smart home has doors that can open electronically, lights that can turn on using voice command, temperature control, interconnected appliances, and security features that link to a smartphone.

A smart home can have a hundred different features integrated, which is why many IoT developers are very interested in this field. For these devices to operate seamlessly, they require 5G technology for its speed and data capacity.

Smart homes can also come with AI voice programs designed to provide both friendly chitchat and basic services like shopping, playing music, providing sports results, making to-do lists, and welcoming guests.

Even ordinary homes built without smart features can have them integrated today, which is precisely what South African firms Touch Technologies and Harrison Homes does. This company offers services ranging from wireless light control and CCTV installation to temperature control and voice-activated entertainment systems.

New AI-powered homes can even study your moods and facial expressions to change the colors, patterns, and ambiance of your home to match your mood or to cure the blues. The future of real estate in Africa is in technology. Developers who fail to integrate 5G and AI into their operation won't be able to remain competitive.

Conclusion

Any business that wants to invest in the future of Africa in any of the following sectors:

- Education
- Health Care Delivery
- Transportation & Logistics
- Hospitality (Hotels, Restaurants, Bars)
- Manufacturing
- IT Services including SaaS
- Telecommunications
- Agriculture and Food Processing
- Banking & Financial Services
- Fashion & Design
- Real Estate Development

Needs to embrace, integrate, and use 5G and AI to remain competitive and ensure that technology helps Africa overcome some of its most outstanding challenges.

Investopedia notes that the best way to invest in Africa is to through Direct Access by opening a brokerage account with a local bank. Besides that, the Investopia resource on how to invest in Africa notes that Mutual Funds, ETFs, ADRs.

www.ingramcontent.com/pod-product-compliance
Lightning Source LLC
Chambersburg PA
CBHW070436220526
45466CB00004B/1699